SCIENCE EXAMINATIONS

A QUARTLY SERIAL PUBLICATION

copyright 2017 all rights reserved by:

Love From The Sea publishing company and Dr. Ronald Cutburth

ISBN 9781521460023 JUNE - VOLUME 2.

Featured article:

NO NUCLEAR WEAPONS

ONLY EXPLOSIVES DESTROYED

WORLD TRADE CENTER BUILDING ONE

A forensics analysis By:

Dr. Ronald Cutburth, engineering scientist, intelligence expert.

INDEX THIS ISSUE:

Introduction

A. Technical notes on standard nuclear weapon power of existing weapons developed in the 1960s and forward by the US and USSR

B. The 1993 World Trade Center Building One attempted destruction.

C. What the weapons can do/ did not do; nuclear and non-nuclear, explained in simple Mechanical Engineering methods.

D. Explanation of why reported radiation scattering is not evidence that a nuclear weapon was/was not used. Mechanical engineering needs to be used.

E. Claims of ultra-mini-nuke development are not supported by scientific fact or historical rationale.

NO NUCLEAR WEAPONS

ONLY EXPLOSIVES DESTROYED

WORLD TRADE CENTER BUILDING ONE

A Forensics analysis

BY

Dr. Ronald Cutburth, Ph.D., Management of Engineering Science Operations, engineering scientist, intelligence expert.

Ph.D dissertation includes an evaluation of more than 200 research projects accomplished at Lawrence Livermore National Lab. This was the culmination of working experience in engineering projects across America. That work included over 9 years of experience in the engineering checking of engineering designs. This includes military and commercial programs. Below find a forensics analysis of the demolition of WTC-ONE. The North Tower.

INTRODUCTION:

The March issue of this magazine showed evidence that high tech explosives were used to destroy WTC ONE. This issue's primary article is devoted to a forensics analysis providing evidence that no nuclear weapon was used to destroy any part of the building. This analysis includes also a brief comparison of the striking relevance on a few past terrorist demolitions . This presentation shows evidence and calculations not found in any analysis by any who claim nukes were used. We negatively fill that absent calculation and evidence gap here. As many were shocked the building floors were turned into powder, it was easy to be fooled by those who deliberately provide dis-information and those from rumors. The analysis follows.

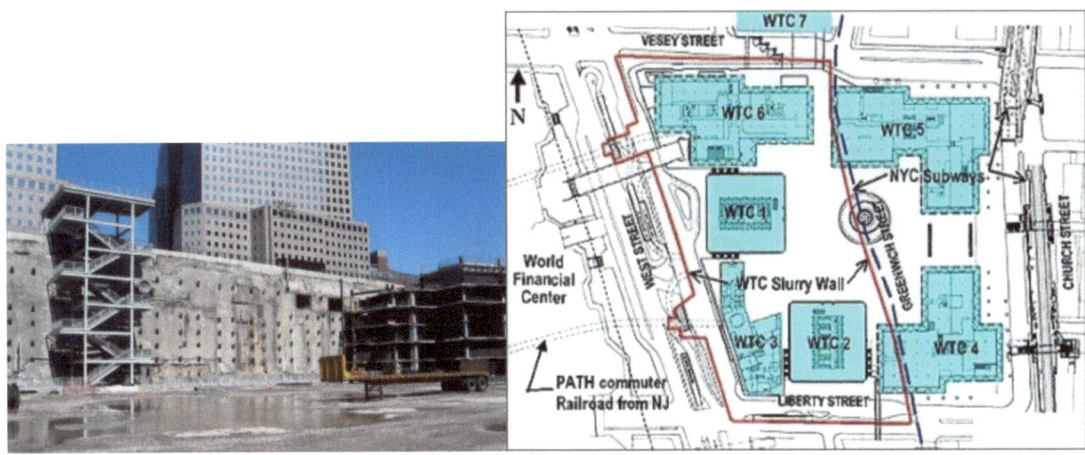

Diagram of the location of WTC buildings and their underground area Slurry wall called the bathtub. We use this Bathtub for analysis below.

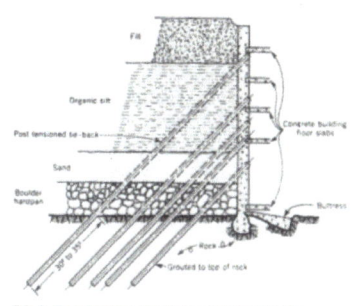

FIG. 3. Typical section shows placement of tie-backs. After concrete floor slabs have cured, tension on tie-backs will be released.

A. Technical notes on standard nuclear weapon power of existing weapons developed in the 1960s and forward by the US

One key issue of the claim of a nuclear weapon is who has access to what weapons. Another issue is the explosive effects on various sizes of weapons as compared to specific effects shown by actual destruction of WTC ONE. The selection here is limited to the range of mini size and mid-range nuclear weapons.

One other aspect is the rumor/myth that ultra mini size weapons can be/were built for this claimed destruction of WTC ONE. The theoretical minimum size of a mini-nuke is about 11-15kg. . Davy Crocket mini-nuke, below has an estimate range of .o1-.02 kiloton of TNT

Free information: http://en.wikipedia.org/wiki/Tactical_nuclear_weapon Search by topic.

This US type W33-208 mm self-propelled artillary is with about 5 kiloton Yield whereas the Soviet counter is for a 152mm with about 1 killoton yield. We compare these in the analysis below. The US version was mounted in small rockets as well. They were given to NATO countries making them more accesssible. Davy Crocket mini nuke, below has only an estimate range of .o1-.03 kiloton of TNT

The carrying case for the W54 Special Atomic Demolition Munition (SADM). The SADM had a yield of 0.01, or 0.02-1 kiloton and was operationally deployed between 1964 and 1988. The entire unit weighed less than 163 pounds (74 kilograms). This is the claimed suit case size nuclear weapon. We examine large nuclear weapons below. There is a multitude of sources with the claimed minimum size of a mini-nuke is around 11-15 kg. They all explode in a round ball. As that minimum size produces the recorded low of .01 kiloton of TNT, this would be much to massive for controlled demolition. This would not be suitable to use as a controlled demolition device. Those must be in shaped charges requiring a non-round explosion.

MEDIUM SIZE NUKE

Medium Atomic Demolition Munition (MADM), shown above, was a tactical nuclear weapon developed by the United States during the Cold War. They were designed to be used as nuclear land mines and for other tactical purposes, with a relatively low explosive yield from a W45 warhead, between 1 and 15 kilotons. Each MADM weighed less than 400 lb (181 kg) total. They were deployed between 1965 and 1986.[1]

http://web.archive.org/web/20060318020109/http://www.brook.edu/FP/projects/nucwcost/madm.ht Brookings institute report.

The carrying case for the W54 Special Atomic Demolition Munition (SADM). The SADM had a yield of 0.01, or 0.02-1 kiloton and was operationally deployed between 1964 and 1988. The entire unit weighed less than 163 pounds (74 kilograms).

Explosive strength measured results in Nevada underground nuclear weapons tests.

The Containment of Underground Nuclear Explosions

October 1989

NTIS order #PB90-156183

After an underground test, several minutes to days later, once the heat dissipates enough, the steam condenses, and the pressure in the cavity falls below the level needed to support the overburden, the rock above the void falls into the cavity as shown in the sketch below. Depending on various factors, including the yield and characteristics of the burial, this collapse may extend to the surface. If it does, a subsidence crater is created.[26] Such a crater is usually bowl-shaped, and ranges in size from a few tens of meters to over a kilometer in diameter.[26] At the Nevada Test Site, 95 percent of tests conducted at a scaled depth of burial (SDOB) of less than 150 caused surface collapse, compared with about half of tests conducted at a SDOB of less than 180.[26] The radius r (in feet) of the cavity is proportional to the cube root of the yield y (in kilotons), an 8 kiloton explosion will create a cavity with radius of 110 feet.[28] r= 55 times the cube root of y. (per wikipedia notes on US underground tests)

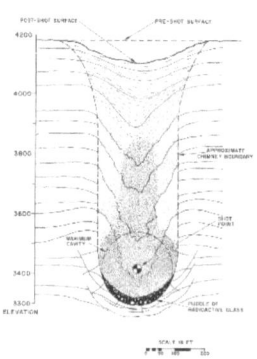

Above is a sketch of an underground test showing the chimney boundary (the blast maximum height) which is well under the surface causing a collapse of the surface to fill the void. The underground tests start with a small bore hole or tunnel that gets expanded into the diameter area that consumes most of the energy of the blast. The resultant upward chimney radically diminishes the remanding upward force. Re-stated the blast force is mostly distributed in the ball area shown in the sketch. That expanded cavity is found in the formula list below.

We see that in the WTC ONE underground garage area, the blast is instead distributed mostly below- in the vast garage area, (the Bathtub) as opposed to making a chimney of force.

Zones in surrounding underground rock. The expanded cavity caused by various size weapons in the kiloton (thousand tons of TNT range).

Name Radius

Melt cavity…………………………… 4 – 12 m/kt1/3

Crushed zone ………………………30 – 40 m/kt1/3

Cracked zone ………………………..80 – 120 m/kt1/3

Zone of irreversible strain ……800 – 1100 m/kt1/3

Blast energy is distributed as Blast= 50%, Thermal energy=35%, and fallout radiation = 10%.

The calculations below are extrapolated from above underground calculations from https://en.wikipedia.org/wiki/Nuclear_weapon_yield. They provide calculations and references not detailed here. Also we note that the surface area of the explosion follows the round ball calculation of a sphere where Volume =(V = (diameter x pie)^3/6) Our destructive possibilities are derived from this calculation of the sphere volume-against a rock surrounding.

In contrast, the blast strength underground in our Bathtub is predominantly toward concrete and steel destruction in an very large open space as opposed to rock of an underground test. **The key difference is that the Bathtub is essentially an underground open space-not a space that is cause by compression of rock. The underground space would cause a loose spreading of the explosion to a much wider area than shown in the list above.**

Due to this open space Bathtub we can estimate direct damage to the building columns and parking levels and building ground floor (no ground floor damage is shown). A key difference in an underground test by the US is the vast space of the Bathtub would allow energy dispersion. It would spread its radiation to a wider area and absorb the radiation as in an underground test.

Historically the common error is to claim a nuclear weapon was used like an above ground test, as opposed to the actual relationship of the underground open Bathtub space. In fact the Bathtub open space acts as a large dispersant of the weapon energy as shown below. The Bathtub open space would also absorb most of the claimed radiation.

Historically arguments have been presented of radiation found around the area in the dust. This analysis shows a fallacy in that analysis.

As the radiation would be mostly held in the Bathtub, the radiation claims from radiation found in the dust would thus not be measured accurately. Also there was some another source of radiation other than a nuke. This would be from nuclear materials stored in the offices of the CIA previously located in WTC SEVEN and totally demolished. This would scatter radiation with the dust clouds.

http://www.nuclearweaponarchive.org/Usa/Tests/Nevada.html Los Alamos official list of nuclear weapons tests. Most of the list capacities are much larger than would be used for one building.

The above photo on the left is of an underground containment failure. Massive, prompt, uncontrolled releases of fishion products, driven by the pressure of steam or gas, are known as venting: An example of such failure is the Bandberry test. Slow, low- pressure uncontrolled releases of radioactivity are known as seeps: <u>These have little to no energy</u>. Re-stated- They have no destructive energy.

https://en.wikipedia.org/wiki/Underground_nuclear_weapons_testing

The photo on the right above shows a small about a one KT, one thousand ton TNT equivalent blast above ground nuclear test. Its power is expended in one millisecond in a round ball. It shows a blast radius and a smoke stack up to a mushroom cloud. The mushroom cloud also has almost no force, in the order of one pound per square inch. As the explosive power is exerted in all directions, the power blowing to earth, makes a cavity and the remaining power joins the upward force making the chimney. They then they slow to a make the mushroom cloud. As the mushroom cloud has no power, it cannot be assumed that any mushroom cloud can reach up and destroy the WTC upper floors. Re-stating; when the power is exerted in a round ball near earth, that is what would happen in all nuclear weapon blasts in one millisecond.

B. The 1993 World Trade Center Building One: attempted destruction.

" The bomb exploded in the underground garage, generating an estimated pressure of 150,000 psi.[11] The bomb opened a 30-m (98 ft) wide hole through four sublevels of concrete. The detonation velocity of this bomb was about 15,000 ft/s (4.5 km/s). Initial news reports indicated a main transformer may have blown, before it became clear that a bomb had exploded in the basement.

The bomb instantly cut off the World Trade Center's main electrical power line, knocking out the emergency lighting system. The bomb caused smoke to rise to the 93rd floor of both towers, including through the stairwells which were not pressurized, and smoke went up the damaged elevators in the World Trade Center Towers 1 & 2. ; With thick smoke filling the stairwells, evacuation was difficult for building occupants and led to many smoke inhalation injuries. Hundreds were trapped in elevators in the towers when the power was cut, including a group of 17 kindergartners, on their way down from the South Tower observation deck, who were trapped between the 35th and 36th floors for five hours.

Yousef was assisted by Iraqi bomb maker Abdul Rahman Yasin, who helped assemble the complex 1,310-pound (590 kg) bomb, which was made of a urea nitrate main charge with aluminum, magnesium and ferric oxide particles surrounding the explosive. The charge used nitroglycerine, ammonium nitrate dynamite, smokeless powder and fuse as booster explosives.[18] Three tanks of bottled hydrogen were also placed in a circular configuration around the main charge, to enhance the fireball and after-burn of the solid metal particles.[19] The use of compressed gas cylinders in this type of attack closely resembles the 1983 Beirut barracks bombing 10 years earlier. Both of these attacks used compressed gas cylinders to create fuel-air and thermobaric bombs[20] that release more energy than conventional high explosives. According to testimony in the bomb trial, only once before the 1993 attack had the FBI recorded a bomb that used urea nitrate. The Ryder van used in the bombing had 295 cubic feet (8.4 m3) of space, which would hold up to 2,000 pounds (910 kg) of explosives. However, the van was not filled to capacity. Yousef used four 20 ft (6 m) long fuses, all covered in surgical tubing. Yasin calculated that the fuse would trigger the bomb in twelve minutes after he had used a cigarette lighter to light the fuse.

The bomb exploded in the underground garage, generating an estimated pressure of 150,000 psi.[11] The bomb opened a 30-m (98 ft) wide hole through four sublevels of concrete. The detonation velocity of this bomb was about 15,000 ft/s (4.5 km/s). Initial news reports indicated a main transformer may have blown, before it became clear that a bomb had exploded in the basement. Yousef was assisted by Iraqi bomb maker Abdul Rahman Yasin, who helped assemble the complex 1,310-pound (590 kg) bomb, which was made of a urea nitrate main charge with aluminum, magnesium and ferric oxide particles surrounding the explosive. The charge used nitroglycerine, ammonium nitrate dynamite, smokeless powder and fuse as booster explosives.[18] Three tanks of bottled hydrogen were also placed in a circular configuration around the main charge, to enhance the fireball and after burn of the solid metal particles.[19] The use of compressed gas cylinders. "

Above outline is mostly quoted comments found at wikipedia.

In this type of attack it closely resembles the 1983 Beirut barracks bombing 10 years earlier. Both of these attacks used compressed gas cylinders thermobaric bomb to create fuel-air and thermobaric bombs that release more energy than conventional high explosives. According to testimony in the bomb trial, only once before the 1993 attack had been used per the FBI testimony. There was one other 1983 terrorist attack that was on the US embassy building also in Beirut Lebanon. In details there was a claimed bomber who made a telephone call claiming the terrorist attack. By this anonymous call the Iranian government was blamed. The Iranians are predominantly Shea Muslim. That fact was ignored in the fine of Iran for the barrack attack- by extension of one anonymous call on the embassy attack.

The 1983 barrack attacks were of the same technology used in the 1993 WTC bombing. Those terrorists were Sunni Muslim. This author thus shows a systemic failure in the blame of Iran (Shea Muslim) versus perhaps Saudi Arabia (Sunni Muslim). Jumping forward the WTC buildings are blamed on Saudi Arabia (Sunni Muslim) not Iran (Shea Muslim). At this writing one can see that the technology used on both the barracks bombing and the 1993 WTC bombing is the same as stated in testimony by the FBI.

Furthermore the evidence for the embassy bombing was by means of an anonymous call. An anonymous call on a trial for murder would not provide evidence in a case of murder. In contrast we have convictions of the Sunni Muslims in WTC ONE who used the same technology in the 1993 attack on WTC ONE as used in the Marine Barracks bombing. This seems to provide

evidence of a method of operations by the global Sunni Muslim leadership. Furthermore the Saudi Arabia (Sunny Muslim) country are the claimed 911 demolishers of WTC ONE.

The Same method of operations provided by Sunni Muslim convictions points to Sunni Muslims as the only suspects for both the US embassy bombing and the US Marine barracks bombing killing more than 220 Marines, and the 911 Sunni Muslims 911 attack on WTC ONE. Certainly an anonymous call did not provide evidence against Shea Muslims.

As the 1993 bombing were by Al-cheda connected terrorists and suggested Osama Ben Laden. The Al-cheada in previous history were formed by the CIA to fight the Soviet Forces in Afganistan. Thus one should ask why the CIA was not questioned on the 1993 WTC bombings and on the Marine Barracks bombing. Or perhaps also the 911 WTC bombings.

In contrast the FBI brand new director immediately blamed Osama Ben Laden and the Saudi Arabia claimed pilots. This was an immediate assumption of Sunni Muslim guilt. The 1993 FBI investigation of the explosives was thorough, but history shows there was no FBI investigation of the 911 WTC bombings for explosives. Explosives were found in the dust by many citizens. Below find photos of the 1993 WTC ONE damage.

In comparison of the explosive force shown by this special high explosive shows us facts. It shows the force in the order of the mini nuke shown above as the Davy Crocket nuke. Shown below then is the facts that refute the claim of a mini nuke destroyed the WTC ONE is erroneous, done without research and simply rumors. Or if it included CIA help the CIA or FBI, the plan appears to blunt their plans. Furthermore the terrorists claimed they planned to cause WTC ONE to fall over and hit WTC TWO collapsing both.

Except this 1993 bombing test provided the facts that a mini nuke would not bring the buildings down. For the 911 destruction they used high explosives on every floor, as shown in the magazine article "EXPLOSIVE DEMOLITION OF WTC BUILDING ONE, A forensics analysis, By Dr. Ronald Cutburth" found at Amazon.

For each floor an estimate of 10 tons TNT (.01kg)equivalent in high explosives were used for each floor noted above to destroy only one floor of the 110 floors This accumulates at 110 floors accumulates to about 1100 ton of TNT equivalent to destroy all the floors.

A rough sketch of the destruction is shown next drawn for public use. For this report this author classifies this 1993 destruction as a "pro-tem" high tech "thermobaric" bomb that included materials that mimic the explosive "Thermite" . This underground test used to test weapons for destruction of WTC. It included high tech elements. As it failed it showed they needed to choose an alternative method of WTC demolition. It provides evidence they had some source of high tech explosives, likely not of their prior knowledge.

1. The bomb was placed on parking level B-1, two levels below the building lobby floor.

2. Notice the size of the holes blown in the garage floors at different layers. The simple fact is the holes in the garage floors get larger but downward they get larger, as the explosive distributes its energy. This demonstrates the explosive primary force went downward-and outward.

3. There is a much smaller hole in the floor above the bomb and a small amount of damage (small hole) in the building ground floor. However that hole is in the weakest part of the ground floor. This shows the ground floor lobby has stronger construction than the parking garage floors.

4. This 1993 demolition is on the order of a mini-nuke. It demonstrates the same function as a mini-nuke and explodes in a round ball. This sketch above shows the round ball would blow holes in the floor but not turn all of the floors into powder as shown in the cover photo. This provides evidence that the claim of use of mini-nukes on every floor is a failed claim because all of the concrete on all floors were turned into powder. Mini-nukes would leave thousands of large chunks of concrete at the bottom, but they are non-existent.

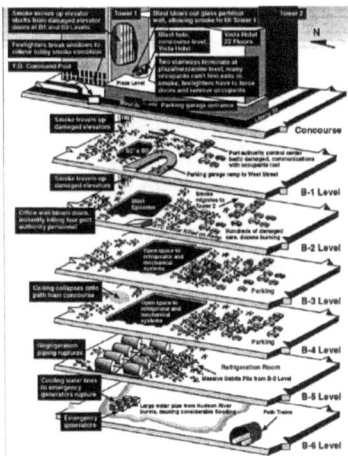
From wikipedia 1993 report

C. What the weapons can do/ did not do; nuclear and non-nuclear, explained/estimated in simple Mechanical engineering technology.

Above was shown the facts on the 1993 attempted WTC ONE demolition. What was tested was a mini nuke sized high tech non-nuclear weapon. It failed.

Below we estimate an explosive power of a nuclear weapon and its estimated damage. Especially the damage pattern as it directly applies to an unproven claimed WTC ONE underground explosion.

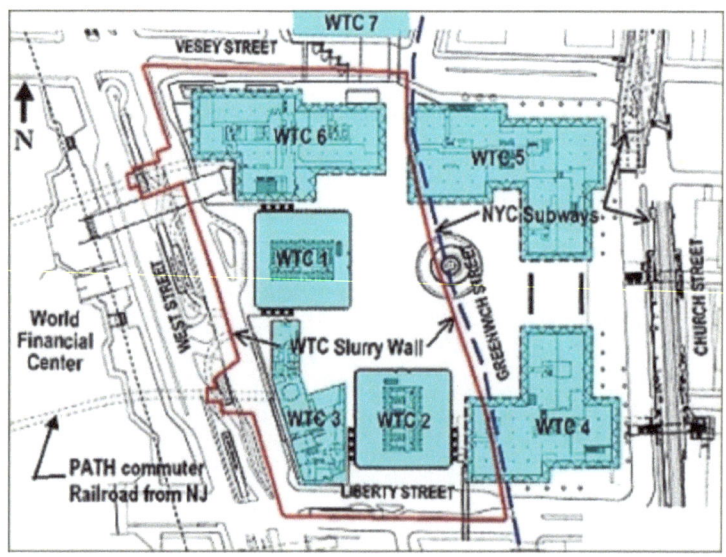

Shown is, outlined in red; The Slurry Wall called the Bathtub.

Below find two photos showing the Bathtub walls.

Above the WTC ONE is almost centrally located in the Bathtub but about 30 feet off center side to side. Some key historical facts exist. What we know by historical facts. Facts list shown below.

1. The Bathtub walls were not destroyed in the 911 attack.

2. Both WTC ONE, and TWO remained standing structurally all during their demolition/collapse-demonstrating no significant bathtub damage of the core columns.

3. The street surface and the world trade center buildings ground floors were not destroyed. This is shown in many videos of the people escaping from WTC ONE and TWO at street and ground floor levels. Facts exist that firemen exited both buildings at ground level after the building began to collapse from top down. A nuclear weapon claim of a nuclear weapons in underground- in the Bathtub would require that the ground floor lobbies and street level be destroyed first around the building as those are above the Bathtub.

4. There were no big holes in the concrete at the ground floor level. (shown in photos below) This is demonstrated in the debris piled on the ground floor concrete; debris is not resting at the bottom of the Bathtub. In addition, surviving fire fighters were rescued from a stair well above the ground floor level demonstrating again that the lobby ground floors were not destroyed.

5. That the ground floor lobbies were not destroyed demonstrates that no nuclear weapon was used to subsequently reach up past the ground floor to destroy the top floors and progressing sequentially down the building. As a nuclear weapon expends all of its energy within a millisecond/microsecond, it cannot follow that underground destruction by reaching to the top of the building for more destruction. Re-stating, a nuclear weapons uses up its energy in one millisecond

6. The sketch of the WTC ONE 1993 demolition attempt shows most of the garage floors had holes blown in them. This demonstrates the fact noted about on the photo on the right that when the nuclear weapon explodes in a round ball a percent of its energy go downward. In an underground explosion in the Bathtub the weapons would destroy the lower and upper garage levels but expend a percent of its energy downward. This would reduce any energy thought to go upward toward the ground floor, through the ground floor then onward up to the upper floors. Also its total blast time is one millisecond. Thus its time to destruct would be expended in the Bathtub only.

As examined below, a nuclear weapon placed in the garage level would cause the buildings to fall, or the ground floor lobby floors and street level would be destroyed without any of the top floor destruction.

All nuclear weapons explode totally within one millisecond/microsecond; All of its energy is gone in one millisecond/microsecond. WTC ONE was destroyed in 8—10 seconds.

<u>**The destructive power of a nuclear weapon would be all expended before one thousandths of the time of the building downward actual destruction. Once the weapon energy is gone it cannot continue to destroy the building for even one second, as opposed to 8-10 seconds of actual destruction as measured seen and seismic recorded historical facts.**</u>

Sketches and calculations follow of a hypothetical nuclear weapon test in the Bathtub; in the claimed reason for the destruction of WTC ONE.

The Bathtub square feet and depth provides the space for a nuclear explosive distribution area, as opposed to the size of an underground test where the earth is expanded by the explosion, But the Bathtub would still absorbed radiation and fall out, shown in the above calculation chart and notes. We nominally chose the rock crush area using the above formula nuke for a couple choices of nuclear weapon sizes. For this comparison or a demolition in the vast area of the bathtub the energy would not be all consumed. $35 meters/kt^{1/3}$ revised for smaller nuke would be,

Crushed rock radius: 16.2 meters (53.2 feet)/$.1kt^{1/3}$, (rock crush radius of 16.2 meters)

We compare a 100 ton TNT equivalent in the Bathtub explosion to the calculated amount of TNT equivalent to demolish all of the floors into powder and floor steel into tiny chips (more than one thousand tons or (called 1kt) of TNT or high tech nanoRDX/nanoThermit used to demolish all the floors). Re-stated the first calculation estimates the energy of 10% of that used to destroy all floors, and is calculated on a radius that matches the radius of 16.2 meters of an underground-under rock demolition to provide a comparative start point.

Continuing rock crush radius calculation and

conversions to foot pounds of force per square inch of ball surface; as all nuclear weapons explode into a big round ball shape.

This calculation uses a start point radius that would be produced by an underground explosion using a 100 ton TNT equivalent bomb. We apply that as a radius for a calculation of an underground explosion in the bathtub that is not limited by the underground crushed rock cavity- giving a nominal start point comparison. In contrast to the underground test used, the Bathtub allows a massive spreading area for its remaining explosive power. Shown below:

Rock crush radius for a 100 ton bomb = 16,2 meters = 53.2 feet gives 106.4 ft diameter,

thus the volume is (pie x 106.4 ^3) /6 = V= V= 630 thousand cubic ft.

Surface area (A) of ball= A= 106.4^2 x pie = A =35.6 thousand ft^2 and as one square

foot equals 144 innches^2 inches is 144 x35,6 ft^2. It follows that this sum is:

35.6 thousand x 144 = A = 5.1 million inches^2. These are in round numbers for simple

application.

When one ton of TNT produces about 3 trillion foot pounds of force we have 100 x3 trillion = 300 trillion foot pounds divided over the surface of 5.1 million square inches gives 300 trillion foot pound /5.1 million inches^2 = about 58 million foot pounds of force per inch^2. As the blast is air much of it rushes around the column at hyper-sonic speed. Total damage to the inner ring of the core columns would be accomplished

However the report above given of the energy, 50% goes to blast, 35% goes to heat and about 10% goes to radiation elements. Thus the load against the core columns is about = about 29 million pounds per inch^2. An estimate of the various core column stability is shown below. It is known that the 24 primary outer ring core columns plus a large internal croup, noted by the photo below are made of exceptionally massive structures. Comparing the explosion ball when it grows instantly to 500 ft diameter the energy is reduced from 29 to about 6 million lb^2.

An estimate of column structural support by the garage floors

Follows in the Bathtub:

The building design places the core columns on the bottom of the Bathtub. The columns contribute to the support of all of the garage floors. The garage floors, that surround the core columns must support autos and small trucks. The floor strength must be rated to carry at least three times the load of trucks.

A line of small trucks loaded with shipping supplies is chosen. We can estimate those weight 6 tons (12 thousand pounds) each loaded. The rated load then must be in the order of 12 tons per 10 feet of floor. When the building is 208 feet wide the floor capacity must be in the area of 208X4= 832lineal feet. At 10 feet per truck this would be 83 trucks at 24,000 lb each =5.76 million pounds per floor. For a 3 times safety factor the floor strength grows to 17.28 million pounds. 7 floors gives us an estimate of more than 120 million pounds of floor strength.

That backup floor strength would multiply because of the rating of the support steel in the concrete and the crushing (compression and pull apart) strength of the steel. Conclude that though the force to destroy each core column is high, they are also supported by the 7 levels of garage floors. As the floor beams would be pulled apart and crushed they would give additional support to the core columns. In a 100 ton TNT equivalent nuclear test estimated above. The Core columns would certainly significantly bend, they may not be torn out altogether.

In the 300 ton TNT equivalent test they would be totally destroyed along with the lower level of the building and street level concrete. Conclude there is no evidence of any size nuclear weapon being used. A 300 TNT equivalent detonation would still be only 30% of the amount of energy to destroy all of the floors. In all cases there is no evidence of ground floor destruction or street level destruction or damage.

Bathtub area, depth, and total cubic feet.

Square feet area: about 480 thousand square feet.

Cubic feet: At 80 ft depth produces an approximate 38.4 million cubic feet.

Comparing bathtub volume verses nuclear weapon explosion volume

When the 100 ton of TNT nuclear weapon is damaging/destroying the core columns it's at a volume diameter of 106.4 feet producing 623.6 thousand cubic feet, it has only consumed less than 1% of the Bathtub volume. There is still more than 37.3 million cubic feet of the Bathtub to expend its energy. This massive energy dispersion area would limit significant damage to the street level in a 100 ton TNT equivalent weapon. In addition there are large air vents. The vast area would absorb more of the radiation and fall out.

In sharp contrast, the amount of force to destroy all of the floors is estimated at more than 1000 tons would produce 3.3 x 58 million foot pounds of force or more than 191 million foot pounds per inch^2. This would destroy all of the core columns and all of the lower part of the building within its millisecond demolition time. The street level around the building would have blown up and out. The building would simply fall over.

Without showing detail, one should accept that when 191 million foot pounds per inch^2 air shock hits the core columns they would substantially bend. As The core columns in the bathtub would be more than 70 feet tall and 22 or 52 inches wide the air blast impact would be 70 (870 inches x 22) providing the sum of more than 18 thousand inches^2 times 191 million pounds per inch ^ 2 produces more than 2 trillion foot pounds of force on one core column. Conclude they would bend like a noodle. Of course when the air blast hits the exposed flat side of 52 inches it would receive more than twice that force. As shown in photos below the core columns show no sign of bending-zero bending.

In contrast the 1000 tons of TNT equivalent nuclear material that was needed to destroy all of the floors compared to 100 ton TNT in the bathtub destroying the core columns, there would be no energy to reach up and destroy the floors; Yet the core columns show no damage. The photos also show no holes in the ground floor. This demonstrates it did not also reach up and destroy all the floors that required more than a 1000 ton TNT group of explosives. Thus the 300 trillion ft lb of force in a diameter is expended in the Bathtub and not in the building. The demolition would be within one millisecond. This provides evidence the claim of a nuclear weapon in the Bathtub to destroy the building from top down is erroneous/significantly flawed. In addition, the size of the weapon to destroy any outer ring core columns is much larger than a mini-nuke.

Primary core columns are examined for their possible damage or destruction below.

As shown in the above analysis a 100 ton TNT rated nuclear weapon would produce hundreds of more than 58 million of foot pounds of pressure per inch ^2. That would at minimum bend all the outer ring of heavy core columns. That bending would show bent core columns coming from the lower level.

As seen in the photos below the core columns protruding from the lower level are exactly straight with no bending. Also the workers are walking on the junk laying on the ground floor. Thus the ground floor was not destroyed. Conclude no nuclear weapon was used.

The photo below is of a long piece of heavy core column. This is one of 24 in the outer ring of core columns that were in both WTC ONE and TWO. A 100 ton TNT rated nuclear explosion would produce hundreds of millions of foot pounds per inch ^2 of force bending all of the large core columns. There would be no straight core columns because of the massive instantaneous force. As the core columns show no bending, one onclude there was no nuclear weapon used.

The image above is from the documentary "Up From Zero" This heavy core column is shown at the WTC 911 museum. The photo shows the base of a core column. This and all outer core columns were mounted on the steel and bedrock at the bottom of the Bathtub. Its dimensions, minus the four flanges, are approximately 52 by 22 inches, with walls at least 5 inches thick. It has no distortions or twisting. It was not as others were not subjected to a nuclear weapon blast. All of the building structures above the ground floor were destroyed, and none below the ground floor were destroyed by a nuclear weapon, or this one would be also destroyed.

D. Explanation of why radiation scattering in the dust of the building debris is not evidence that a nuclear weapon was used. No nuclear weapon was used based on facts.

Claims and counterclaims about radiation found in the dust surrounding the WTC demolitions have been published that are not based on a valid statistical analysis source.

1. For the ones claiming a nuclear weapon was detonated in the underground (bathtub) they supply no mechanical engineering calculations in support of their claim. The mechanical engineering facts shown above demonstrate there was no nuclear detonation in the Bathtub.

2. Historical facts are available for the US underground nuclear tests done in the 1960s. The essential purpose of the underground tests was to contain the radiation and stop

the subsequent fall out. As this analysis demonstrates that in a 100 ton TNT equivalent blast that would heavily damage the core columns, no damage is show, thus that radiation would have mostly been contained in the Bathtub. This demonstrates the radiation that might have been found in the dust can not be in sufficient quantities to project a statistical analysis. Thus the claimed statistics would not be invalid.

3. This author evaluated scientific programs by Lawrence Livermore National Lab. One included their development of a small portable nuclear material evaluation unit. This can be distributed in all CIA locations that may need to test nuclear material. One would be found in the CIA offices found in WTC 7. Along with this would be comparative nuclear material. This fact would account for some of the nuclear samples found in the dust. The CIA would never admit they held sample nuclear material in the City of New York. Thus radiation material found in the dust would not represent a nuke was used.

Conclusion on claimed nuclear weapons use: none was used.

In this analysis, the examination of a 100 ton TNT equivalent blast in the Bathtub shows that the core columns would have been significantly damaged, but no damage is

shown, demonstrating that no nuclear weapon was used. This also puts to rest that a small mini-nuke could have been used that has less than 5 % of the power of this nuclear weapon analysis.

The claimed nuclear weapon in the bathtub also includes the claim that the nuclear weapon also reached up and destroyed all the building floors. The evidence above shows the ground floors were not penetrated for that to happen. In addition the amount of explosive material that was required to demolish all of the floors (330 TNT equivalent) would destroy the entire bottom of the building and not have blast time, of one millisecond, to reach to the top. All photos and videos show the building was destroyed from the top down. In addition no damage is show to the core columns in the Bathtub.

Shown in the photo below are many fire fighters standing in the area of the building ground floor. The photo shows there is no destruction of the ground floor, as would be required for a nuclear weapon to reach up and start destroying the building floors from the top down. The destroyed ground floor would also represent the destruction of all of the core columns and parking levels below the ground floor. This demonstrates again that no nuclear weapon was used.

E. Claims of ultra-mini-nuke development are not supported by scientific fact or historical rationale.

1. For a nuclear weapon chain reaction to detonate, its nucleus requires it be first detonated by a high explosive of some kind with sufficient volume and force to start the nuclear chain reaction. When attempting to drop down to a small and smaller size the nuclear material must always be surrounded with a sufficient quantity of high explosive. The danger is thus presented that there must be sufficient nuclear material that sustains a continuous and growing chain reaction. This combination produces a limitation of the minimum size of a mini-nuke. The reported size is no less than 10-11 kilogram. Adding the high explosive and the detonation mechanism grows the size. Its thought to be just below the size of the mini-nuke called the Davy Crocket shown above.

2. As high explosives are needed as well as safety container and detonations mechanism, the required space consumed renders it impractical for simple explosives. Or simply the high explosives of today offer more safety and better control. Where the mini-nuke must be in the shape similar to a round ball, it is more bulky; thus more difficult to conceal. We have seen the simple flat shaped charges the suicide terrorists wear.

3. For the claimed use of mini-nukes on the WTC ONE, we find that a mini nuke size weapon is too small to do significant damage. There are thoughts of the use of mini-nukes use to demolish the floors. That simply does not apply since all mini-nukes detonate in a round ball. This cannot suggest application on a 43 thousand square foot building floor. Any mini-nuke arrangement would instead blow big holes in the floor and not destroy its supporting steel. It would also leave large chunks of concrete. The WTC ONE demolition showed that all of its floor concrete on every floor was evenly and totally turned into powder.

4. As all nuclear explosions explode/detonate in one millisecond, for this some of a distributed number of mini-nukes would likely not detonate. This would provide evidence of controlled demolition using mini-nukes. Using high tech high explosives ensured even detonation and thought to be-no evidence. A forensics, criminal analysis was required to expose it.

5. Neutron bombs will not work to demolish buildings made of metals. One claimant suggested evidence of a mini- neutron bomb was place in the Bathtub and the neutrons shot up and melted the tall tower on top of WTC ONE. Certainly there is no explanation by anyone claiming and proving a nuclear weapon was used. This provides more evidence the proponents of the nuclear weapon claim do no research. In conflict is physics. The Neutron derives its name because the elementary particle called the neutron is thus called because it has no charge-it is neutral so that there is no charge, as compared to a proton or electron it has no charge. Thus it penetrates right through metals. It will not be absorbed, thus melts nothing. They are left with no explanation of the demolition of the tower and its massive support structure.

Historically the US claimed their scientists developed a neutron bomb. The claim shown in history is that it can be used in a tank battle and when detonated would kill the tank operators without destroying the tank. There is no evidence the bomb was actually built and it is stated that they chose to not build it. Thus there is no evidence of what amount of neutrons it would produce. The sad thing is it would likely kill more civilians than military. It would have no effect toward the destruction of WTC ONE.

Copyright 2017 by Dr. Ronald Cutburth and Love From the Sea publishing company

Below find more photos of ground level with no hole in it from a nuclear weapon, and another view of the Bathtub showing it was not destroyed.

Below find a photo of an passenger plane destroyed by hitting a simple bridge railing.

Notice its structure is flimsy-ok for flying only. Could not destroy WTC ONE.

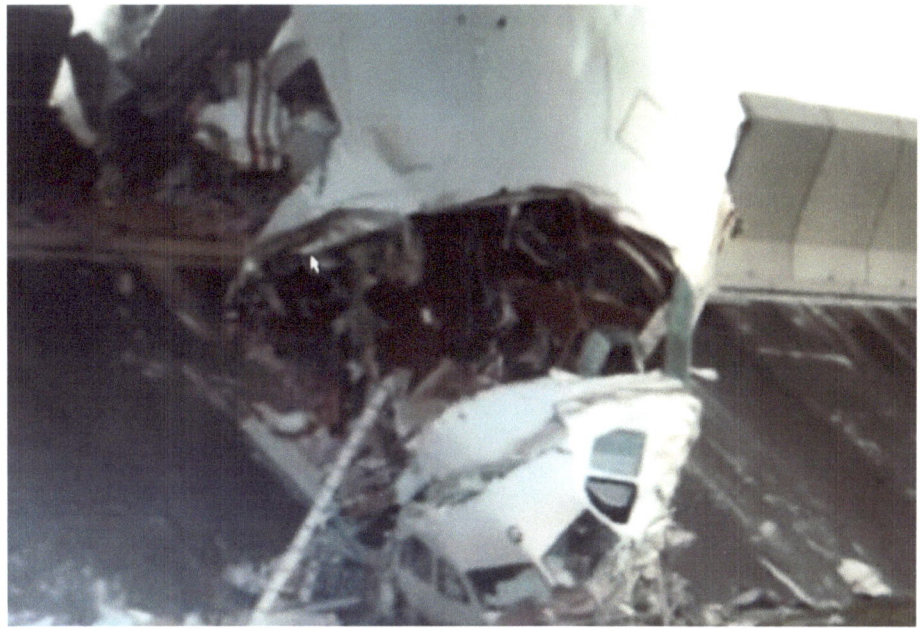

www.ingramcontent.com/pod-product-compliance
Lightning Source LLC
Chambersburg PA
CBHW041307180526
45172CB00003B/1011